10,000 Birds
Volume I

Written and Illustrated by
Christina Baal

Mama - Ryan - Allison - Grammy - Grandpa - Aunt Colleen - Aunt Siobhan - Uncle Ken - Cutie - Snuggles - Achilles - Samara - Slim - Clementine - Susan Fox Rogers - Lisa Sanditz - Kenji Fuijta - Tatjana - Kate the Kayaker - Kat Jump - Margaux Robles - Sarah Watson - Leila Duman - Anna McKeigue - Sumedha Guha - Emma Pile - Sorel Dunn - Maya McLaughlin - Bard Cross-Country - Erin Kelly - Maddy Lemann - Coe Walker - Christine Nassar - Janell Smith - Kaila Pedersen - Midori Takasaki - John Burroughs Natural History Society - Ralph T. Waterman Bird Club - Peter Schoenberger - Mark DeDea - Everyone who supports me and 10,000 Birds

Thank You.

This book is dedicated to you. Everything you do means the world to me.

Special thanks to Maddy Lemann and Erin Kelly
for helping me make this book a reality!

I have loved drawing animals ever since I could hold a pencil. When I was younger, I would copy pictures from animal books, living vicariously through the experiences of others. But now, by looking for birds, I am the person having adventures and getting to see the most incredible wildlife and landscapes. More than that, though, by being able to see these things and meet each bird firsthand, I have discovered that every one of them has its own quirky personality and lifestyle. It is in these moments when the threads of our lives intertwine- those wondrous moments when my heart quickens as we coexist in this world together- I learn to see each bird for more than just what my own prejudices would have me see. And it is in these moments that I also learn something about myself.

My ultimate goal is to see and draw all 10,000 species of birds in the world. Such a quest will bring me to places I could never even imagine and let me see things that will take my breath away. By the time I have seen them all, I will have visited every continent, traveled thousands of miles, encountered all sorts of wildlife, seen unspeakable beauty and met all sorts of people from all different walks of life.

One of the most remarkable things that I have discovered through birding is that birds are a universal language. No matter where I go, when people see me with my binoculars they want to talk to me about the birds they have seen or hear about my adventures. I have met so many people through birding. Countless numbers come up to me to ask if we can take a walk together to look for birds. These winged creatures allow me to have conversations with people I have never met before because birds are something the whole world can share. Therefore, I truly hope that my journey to find all of the birds of the world, and the art that comes out of it, will be the way in which I can connect with people to show them the awesomeness of the natural world, that we are irrevocably a part of it, and, ultimately, that we must work together with each other and the living creatures we live alongside to protect it.

"You see what you love." Such a simple idea, really. And this concept, so beautifully described by John Burroughs, is the guiding principle behind 10,000 Birds. When I show people my bird drawings, I am constantly asked the deceptively straightforward question: "Why do you love drawing birds so much?" It is hard to articulate a simple answer. I could reply, from an artist's standpoint, that it is because birds are aesthetically and behaviorally mesmerizing. Or, I might try to explain from the perspective of an athlete. Birding can be competitive: a single species prevents me from retaking my position as the number two birder in Dutchess County. Whenever I receive news that the woman who sent me to slot three has found a new bird that I have yet to meet, I take off on my bicycle to track it down. I do not think I will ever forget when I tried to find the Cliff Swallows she had reported. When I arrived at the exact coordinates twenty-seven miles later, I found a house. I decided my rival must have trespassed onto the property to find the birds that eluded me. It was only later that night I realized she had not been trespassing. Instead of swallows, I had found out where she lives!

Most often, though, I tell people that birds have inspired some of the most awesome adventures I could have dreamed about. Chasing after birds has led me to do things I never imagined I could do, whether it be biking fifty miles in a single afternoon to see Purple Martins, hiking up mountains for eleven hours to find Bicknell's Thrush, or kayaking in a thunderstorm looking for Sora rails and Least Bitterns. As I draw the birds I encounter, I no longer need to rely on copying or living vicariously through the works of others. Instead, I can draw pictures from my own sketches and photographs that recall the moments of discovery. In this way, the birds I draw are special- they are more than just an anatomical rendering meant to bring the bird to life on paper. They are also a story; a synthesis of my memories of the time and place when the threads of my life and the birds I met became intertwined. And it is because I feel connected to them in this way that the true reason I love drawing birds so much *is*, actually, remarkably simple. It is because I love them.

It was only after falling in love with birds that I learned how to start seeing them. I am hoping that by using art and writing to share my experiences and the wondrous birds I have met with others, I can try to entice them to fall in love with birds as well. They are wonderfully companionable creatures, living both alongside and removed from us, so different and yet still oddly like humans. We can learn from them as we appreciate the complexities of their lives. Perhaps then, in this way, we can work to protect them. After all, people want to take care of what they love.

This is the first volume of a series that will take me years to finish. This first step has been a learning experience, and certain formal decisions will surely evolve as the series continues. When I first started drawing each of the birds I have seen so far, I did so haphazardly. Now that I have more of a plan, I am going to start drawing birds in the chronological order that I have met them, so that future volumes can be more streamlined in terms of the encounters and accompanying journals. Eventually, each book will continue where the last one ended so that consecutive volumes will tell a story.

I am so excited to continue this journey. I sincerely hope that you can take something away from this first book, and that you will join me on my quest.

Good birding and happy trails!

Table of Contents

Turkey Vulture

"Cathartes aura"

The remnants of the deer that once pulsed with life are consumed, bit by bit, as the turkey vultures dance around its broken body, plunging their heads into its rotting flesh over and over again. There is something else at work here, something deeper than the ponderings of my human mind and the naturalness of the scene before me. It is a fundamental magic that the vultures unleash as each bite they take synthesizes the energy of the deer with their own. They transmute her body into new life, breaking the cycle of death and heralding the promise of immortality. An end to endings. The deer will run again on the wings of the turkey vulture. The dying sun does not need to worry as it sinks lower in the sky beyond the cornfields; thanks to the turkey vultures, it will be reborn tomorrow. The gruesomeness is at the same instant beautiful-death, life, they are both gruesome, both beautiful- and it is this thought that I cling to as I leave the peace eagles to their alchemy.

July 2, 2012. Bard College, Annandale on Hudson, New York

Thermals (convection currents)

Updrafts caused by the uneven heating of the air near the surface of the earth. The air over cities, bare fields, or rock faces heats more quickly than over forests or bodies of water. Since warm air expands and is lighter, it rises above cool, dense air.

Hot air is pushed up.

HOT

H
O
T

C
O
L
D

COLD

cool air is heavy and sinks

Field

Water

Vultures can fly slowly, which is highly beneficial because it allows them greater maneuverability. A vulture needs to fly in tight circles to take advantage of thermal columns, which are narrow.

By gliding from thermal to thermal, vultures can cover hundreds of miles a day with little energy. They watch each other to figure out where the next thermal is.

Its bald head is useful because there are no feathers to get dirty when it sticks its head into a rotting carcass.

*There must be a thermal over the soccer/paintball field!

Turkey vulture flight pattern

Vultures soar by riding updrafts They exploit two types of air currents:
1) Obstruction currents 2) Thermals

wind

Mountain

OR

wind

mountain   Plain

...standing waves – each has an updraft component!

Obstruction Current

Occurs when a prevailing wind strikes and then rises over an obstacle such as a mountain or building

OR

wind coming down the lee side of a mountain strikes a plain and rebounds in a series of "standing waves" like water flowing over a large rock in a waterfall.

TURKEY VULTURE

TURKEY VULTURES
BARD SOCCER FIELD
JULY 15, 2012

MYTHOLOGY

In Cherokee mythology, the Great Vulture played an important role in shaping the earth. When the earth was new, the Great Vulture flapped its wings, and where the tips struck the ground, they formed mountains and valleys.

Algonquians call the vulture a "peace eagle." It brings an element of peace to a tribe of warriors - it doesn't kill. It only eats animals that are dead.

White-eyed Vireo
"Vireo griseus"

It is a strange thing, to see something amazing and not know what exactly it is you are looking at. Recognition, and identity-these two things can change any sort of interaction. I was missing both when I first found the White-eyed Vireo in the Tivoli Bays. I knew it was an interesting bird, one I had never seen before (but at that point, so was pretty much every bird!) but I had no idea just how special it was. I had seen something incredible without realizing it.

I had sworn to myself that if I ever met my vireo again, it would be much different. I would recognize the bird for what it was, and when I looked upon it I would be able to appreciate it as a White-eyed Vireo and not just a nameless specter. But seeing it the first time had already been an incredible bit of luck. Seeing it again would be nothing short of miraculous.

July 2, 2012. Tivoli Bays, Annandale on Hudson, New York

Gray Catbird
"Dumetella carolinensis"

"I've been hearing this thing that sounds like a dying cat."
"Oh, that would be a catbird."

So I was told when I first inquired about the rather disturbing noise infiltrating my ears. There is nothing quite like walking along in the woods and hearing such a sound following you. This particular cry darts in and out of the bushes, hops up onto branches, dives down into the brush. It is sometimes up ahead, back behind, or just all around-but it never leaves. Once you realize that it is not, actually, a dying cat, it becomes a rather endearing bit of chatter. And the best part of all is that the sound that follows me is embodied by an outgoing, smoky gray bird with a black cap that insists on sharing his complete life story with me for the entirety of our time together. He is an amicable companion, and when I am walking along with his voice around me there is no way in the world I could ever be lonely. At times, he reaches an emotional pitch and puffs himself up with such indignation that I can do nothing but laugh. Relax, I sometimes tell him. This seems to make him even more exasperated, and the frantic narration continues. I admit it; I take pleasure in his woes. But secretly, I think that his whining at times might mask laughter-maybe he is laughing at me, for believing that I am carrying on a substantial conversation with a bird! And so we end up laughing at each other.

July 2, 2012. Tivoli Bays, Annandale on Hudson, New York

Sanderling
"Calidris alba"

The Sanderling is a wavechaser. When a wave breaks, it seizes its chance to rush forward onto the wet sand left in the ocean's wake and forage for invertebrates that may have been washed up onto the beach. The hapless bird must be quick because its time is limited. To the dismay of the Sanderling, another wave soon hurdles towards the shore with a vengeance and the shorebird is forced to race back towards the safety of higher ground. And then, within moments, it is again engrossed in the debris left by the retreating wave. I am sure that this routine is serious business to the Sanderling. After all, it is such a small bird that a good size wave could easily overwhelm it. But in my eyes, the antics of the tiny shorebird are nothing short of a delight. The manner in which the Sanderling runs back and forth, back and forth, over and over again is nothing short of hilarious. I do not understand how anyone could watch one of these wavechasers without laughing. My heart is nothing short of elated from such a whimsical show of naivety. Yet even though the antics of the sanderling are hysterical, they are also admirable. When I try to imagine running headlong into a wave over and over again to grab my food I can realize the bravery of this little bird.

July 27, 2012. The Ocean Club on Smuggler's Beach, South Yarmouth, Massachusetts

CANADIAN GEESE
"BRANTA CANADENSIS"
Spring migration

~The Sounds of Silence~
Wednesday, November 7, 2012

I go for a walk down Cruger Island Road at 4 PM. It is windy and looks like it is going to snow. I do not see anything but a few juncos. Now that there are no leaves, I can see bird nests everywhere. There is one just past the metal gate on the left side of the causeway.

There is a paradoxical loudness that comes with quiet. Even the tiniest movement seems so distinct when it cuts through the silence. I barely brush a branch into a leaf and it sounds like an animal has moved. Whenever something does move- the juncos amidst the autumn debris, the dying plants rustling in the wind, a squirrel exploding as it leaps through the crunchy leaves, the trees groaning as they rub against each other as though trying to keep warm- is a sound of life so solitary it could be the only sound in the world.

Canada Goose
"Branta canadensis"

There is a racket going on at Thompson Pond in Pine Plains. Although at nine o'clock in the morning the sun has been up for barely two hours, the surface of the water is already rippling with intense activity. Hundreds of waterfowl are using the pond as a stopover on their long migration south. The sheer number of bodies packed into one place is overwhelming. Ring-necked ducks, black ducks, mallards, and even a gadwall abound. But the most abundant bird by far is the Canada goose. I have never seen so many geese at once. They are packed so closely together that at times it is hard to tell which long, black neck stems from which dusty brown body. It is the geese that are making the racket with their nonstop, echoing honking. Suddenly, although it does not seem possible, the barrage of honking becomes even louder. Looking up, I see a flock of geese streaking towards us. The long, thin v-shaped lines compress into a black cloud as they wing over my head and towards the pond, circling lower and lower as they prepare to land. As they grow closer, their flight becomes a tornado of recklessness and grace. I watch in wonder as they bank so sharply it looks as though they are rolling in midair, yet they somehow maintain their meticulous trajectory as their long wings create enough air resistance to keep them from careening into the water. When they hit the surface, I realize that I have been holding my breath.

January 1, 2013. Thompson Pond, Pine Plains, New York

Wild Turkey
"Meleagris gallopavo"

Turkeys act fierce. The group in front of me is strutting around as though on a parade, gobbling and squawking like they own the place. The males are certainly not going to let me walk by them without making sure that I know that they are the top dogs of these woods. Their tails fan out, their red gobbles sway, and they puff themselves up until they look like balloons. Even though the whole act is hilarious, it definitely commands respect, and I carefully walk around the perimeter of the Turkey harem as I continue my walk. It suddenly begins to snow. The flakes edge around the skeletal branches of the trees and begin to carpet the forest floor. It is so quiet that the loud gobbling of the Turkeys sounds as loud as a foghorn. I burst out laughing as they come into view. The once-puffed Turkeys are now hunched over and covered in snow as they scrounge for corn around the visitor center feeders. From the warmth inside, I can stare at them through the glass and they do not flinch. I lock eyes with one male, and for a moment I can swear that I can see a crack in his virile stance as the bird is whipped by the cold wind and snow.

January 1, 2013. Marshlands Conservancy, Rye, New York

Christine Isaac

Bald Eagle
"Haliaeetus leucocephalus"

I have seen Bald Eagles before, but they were quick flashes of wings and power that were gone before I could get a good look. Without binoculars, I could not even see their eyes. On this cold, gray March morning, I am waiting for Susan, my professor, to go birding at the edge of the South Tivoli Bay. Suddenly, a chirring, a flash of wings, and somewhere out in front of me a giant shadow is heading towards a tree across the South Bay. It is not hard to follow the track of the eagle, and I watch it wing around and head back toward me. All of a sudden, another shadow hurtles earthward and intercepts the first Bald Eagle. The creamy white head of the adult attacks the chocolate brown of the immature as the two dance in a frenzy for domination. Even though they are combatants, their flight is so graceful it is as though they are fighting a choreographed battle. Ultimately, the adult is victorious. It drives away the young eagle and alights in a tree as though to claim the throne of its dominion. From there, I can look straight into the fierce eyes of a bird whose race fought back from eradication-whose fierce spirit is embodied in both birds that fought today.

March 6, 2013. Montgomery Place, Annandale on Hudson, New York

CANADA GOOSE
"BRANTA CANADENSIS"

Fearless - they were so close I could have bent down and pet them, but they probably would have bitten.

JUVENILE BALD EAGLE
"HALIAEETUS LEUCOCEPHALUS"

Black
stripe
near
eye

yellow
bill

mottled
brown
body

Head mostly white, so
probably almost fully
mature

Killdeer

"Charadrius vociferus"

The Bard Farm has a pair of Killdeer. They establish the farm as their home and look at me with contempt as I attempt to weed the rows that they have staked out as theirs. And then, to my absolute delight, someone discovers a nest. The Killdeer will have babies! If the parents were crazy before, they reach new levels of hysteria as they guard their eggs. The male practically goes ballistic trying to distract potential threats with a fake broken wing. I do not think I have ever laughed so hard while weeding. One evening in May, just as I am finishing my finals, I walk by the farm trying to track down the call of a Ring-necked Pheasant. As usual, I look for the mother on the nest as I walk by- but she is not there. The nest is empty. Fireworks go off in my chest. I rush around the perimeter of the farm looking and listening for the familiar sounds of delirium. I spot the male, and as I scan around him, I notice that small pieces of the ground are moving. This is when the fireworks explode out of control. Its feet are as big as its head. The sight of the baby Killdeer makes me so blissfully happy that I feel as though it is a family member that has been born. They have enormous black legs and feet, but teeny tiny bodies. When they try to move their necks, their whole body jerks. Essentially, they look like pom-poms on sticks. I do not think they could be any more endearing. But then the mother calls to her children, and as they reach her they dive underneath her stomach. I leave the family with the mother nestling her babies

April 5, 2013. Bard College Farm, Annandale on Hudson, New York

Ruby-crowned Kinglet

"Regulus calendula"

It is low tide in the Bays. The mud on Cruger Island Road hungrily tries to assail my feet, but it is not thick enough to stop me. Gurgle, squelch; my feet are released from the mud and then promptly disappear again. The water moving out of the Bays trickles over the gravel. Woodpeckers hammer on trees, sparrows sing, Red-winged Blackbirds caw their hearts out. Oak-a-lee! Oak-a-lee! It is a symphony of sounds, so many sounds that I know, and at the same time so many that fall upon my ears as words spoken from a foreign tongue. I can see small, fragile buds hesitantly protruding from the branches of the bushes bordering the causeway. They herald a promise that spring is, indeed, coming; with it comes an influx of new voices whose owners I have yet to meet. Today, it is the explosive chatter of a Ruby-crowned Kinglet, a tiny bird with an outpouring of energy that far outstrips its size, that becomes my newest acquaintance. It's ruby crown stands out against the brown brambles as the hyperactive bird hops about. Although spring is on its way, today, the Bays are still draped with the décor of winter. Browns and grays are the predominant color scheme of the day. Yet the slits of sunrise that escape through the trees, the small efforts of new plants, and the feathers of birds foretell change. Spring is coming, they say. Spring is coming, and with it, color.

April 10, 2013. Tivoli Bays, Annandale on Hudson, New York

Savannah Sparrow
"Passerculus sandwichensis"

It is another gray day. I wonder when I will wake up to a morning that changes color, and is not merely suspended in smoky clouds. Today, though, I prefer the grayness to the alternative: it is supposed to be raining in torrents, but it is holding off-for now. Susan and I are scouring Greig Farm for sparrows. It is not hard to find them; they shoot up out of the newly planted rows of vegetables like popcorn. Some remain airborne for a few moments, as though curious. Others dart back into the leaves as though on an undercover mission. Either way, it makes getting a good view of the specific markings that separate sparrow species from each other rather difficult. So far, I have only seen Song Sparrows here. But there are other kinds of sparrows, and with patience, Susan and I finally find one that alights on one of the leaves and turns its head in profile. From there, we can see the brilliant yellow eyeline that makes this Savannah Sparrow different from its Song Sparrow companions. But there is something else about it that sets it apart. It is not as forward or plucky as the Song Sparrows. It comports itself more seriously, as though it knows that it is different. Suddenly, I feel water droplets on my hands. Within seconds, the Savannah Sparrow has vanished, and Susan and I are running back to her car as the sky finally opens and releases the rain.

April 12, 2013. Greig Farm, Red Hook, New York

brown crown with black streaks, white stripe in middle of crown

CLAY-COLORED SPARROW
"SPIZELLA PALLIDA"
5.5", 8" wingspan
a twice

pale, clay-gray underparts

Looks just like a chipping sparrow, but the cap was a light streak

Dark feathers with clay edges

Sounds like "who who hoo"

NORTHERN CARDINAL ♂
"CARDINALIS CARDINALIS"

GREAT HORNED OWL
"BUBO VIRGINIANUS"

BALD EAGLE NEST

35

Horned Grebe
" *Podiceps auritus*"

"Look! Look!" It takes me a moment to snap out of my reverie. My "Drawing from Nature" class has been assigned to spend the afternoon drawing the waterfall- a deceptively simple assignment. "Watch the way in which the rocks change the flow of the water. Watch the way in which it speeds up and slows down. Draw its motion." At first, I try to draw the scene before me as a landscape. But as I attempt to convey the movement of the waterfall, my drawing slips away from me. Fighting back frustration, I focus all of my attention on the rushing water. I look for something, anything, that I can use to bring my drawing to life, to convey the power before me. I become absorbed in ripples, spray, tiny funguses that float on the stiller periphery of the water, foam, roaring...and then I hear Amanda yelling. As though waking from a dream, I groggily look around to try to see what she is pointing at. My eyes snap open when I realize that there is a duck-like bird riding the waterfall down towards the swimming hole. I am astounded at its gutsiness. Thankfully, my binoculars are around my neck (even in class) and I scramble down the bank until I am at the water's edge. To my delight, I can tell that it is *not* a duck. The bill is too short. Instead, I find that the gutsy bird is, in fact, a bird with blazing red eyes: a Horned Grebe.

April 17, 2013. Bard College Waterfall, Annandale on Hudson, New York

Baltimore Oriole
" Icterus galbula"

Spring migration is becoming overwhelming. There are so many new birds and songs that I can barely keep up with learning them. Never mind the fact that the tiny buds on the trees have decided to flower, effectively allowing the mystery birds to remain well out of sight if they chose to do so. I am experiencing "bird overload." Thankfully, before I can become hopelessly overwhelmed I manage to pick out an individual song I do not recognize. It is a cheerful tune that sounds almost like a cheer one might hear at a baseball game. I head down the South Bay trail pursuing the song that grows louder as I approach. Finally, I reach a point where the bird must surely be directly overhead. I assume that I will now have to spend the next few minutes straining to catch some sign of movement or color or *anything* to locate this bird in the treetops, but my jaw practically drops out of shock when I look up and see *orange*. The Halloween-colored bird singing the baseball ditty is electric in its vibrancy, and I realize that I am really seeing the color orange as it should be seen.

April 18, 2013. Tivoli Bays, Annandale on Hudson, New York

~Little Ball of Sunshine~
April 21, 2013

Mornings do not begin as gray anymore. By the time I have walked down Cruger Island Road and traversed the trail onto the athletic fields, the sun is shining. It adds a sparkle to the grass wet with dew and lengthy shadows to the trees. If I do not raise my hand directly in front of me, I can barely stand to look East. Inevitably, I will look past my hand to try to see whatever winged thing has just darted across the field in front of me. At this point, the only things I will end up seeing are painful sunspots.

A sweet trilling reaches my ears, and despite the spots still flickering in front of my eyes, I can tell that there is something moving in the brush. I follow its movement as it hops from branch to branch, and finally! It bounces out past the green leaves and I can see it fully.

Now this is a sunspot I am happy to lay eyes on. It is a little male yellow warbler, brilliantly golden, chest flecked with chestnut red. "Sweet sweet, I'm so sweet!" it sings as it bobs about. Yes, you certainly are. Sweet and vibrant, glowing with energy and color. The rest of the green leaves, bright in their own right as they twinkle in the rising sun, become even more vivid as a backdrop to the intense yellow bird. It is hard not to feel cheerful while watching something so bright and sprightly. It is incredible to think that such a tiny thing can harbor the power to bring such sunny spirits!

Montgomery Place, Red Hook, NY

Blue-gray Gnatcatcher
" Polioptila caerulea"

Buzz buzz buzz. A hyperactive orb zips from branch to branch amongst the flowery shrubbery at the edge of Cruger Island Road. The tiny bird is maddeningly fleeting. A snatch of steely gray, then nothing, then another snatch, than nothing. Finally, the Blue-gray Gnatcatcher finds a branch which it seems to like. Its dark unibrow denotes it as a male, and gives the bird that is so chipper in demeanor an annoyed look. It is rather amusing.

April 20, 2013. Vassar College Farm, Poughkeepsie, New York

Lawrence's Warbler

"Vermivora pinus x chrysoptera"

Someone has found a needle in a haystack, so to speak. A bird so rare it has not been seen in the Bays in over fifteen years. To see it would be like finding a Holy Grail. It seems so impossible that it could actually happen, but I have to try. Why not hope for a miracle? The bird we are searching for is a Lawrence's Warbler, a rare recessive offspring of a Blue-winged and Golden-winged Warbler. Such genetic combinations might also produce another hybrid, a Brewster's Warbler. The odds of seeing the former are even slimmer, though, which makes it even more incredible that one has been seen right within the heart of Bardland. The quest for a Holy Grail must be difficult. Every Arthurian legend I have ever read about doing so makes it impossible to argue otherwise. But this little bird is about to defy legend. The bizz-buzz of the blue-winged warbler hints at its presence-could it really be here? And before I know it, a small yellow orb is flitting across the field along Kidd Lane and alighting in a tree no more than fifteen feet from me. Bizz-buzz, I feel my heart contract as I look at the black hood over his eyes and the black bib below his beak. It truly is a Lawrence's Warbler, and it is sitting right in front of me. Oh, how King Arthur would be jealous! Just by deciding to take a walk, to hope, and to be in the right place at the right time, this masked yellow bird has turned an everyday Monday into a day of the miraculous. It seems almost too easy. But who would ever turn down a free miracle?

April 29, 2013. Tivoli Bays, Annandale on Hudson, New York

Wood Thrush
" Hylocichla mustelina"

Every morning, I wake up in darkness and walk down Cruger Island Road. The sun rises behind me, widening the spaces between trees and bringing color to the darkness. The sky becomes dark blue, then purple; the color is serenaded by a chorus of birdsong that grows in intensity with every passing minute. Chickadees, warblers, Song Sparrows, woodpeckers galore. But this morning there is a new sound in the daybreak. It is an ethereal calling, almost instrumental in its smooth rising. "E-o'lay!" The song has a ringing quality, with a tremble at the end. It is simultaneously haunting and beautiful. I search for the source of the singing, but it eludes me. It sounds as though it is coming from lower down in the treeline. I look for movement. A shadow seems to move between branches. Suddenly, a robin-sized bird covered in Sienna brown spots bursts into view. In the quick look I get, I can see the song pouring from its mouth at the same time that I see the vivid white eyering that helps me realize that the mystery morning bird is a Wood Thrush.

April 30, 3013. Tivoli Bays, Annandale on Hudson, New York

Scarlet Tanager

"Piranga olivacea"

A rough-sounding robin causes me to whirl around. When I look up, I realize that I have never really known what it meant to see red before this moment. Blazing against the cerulean sky and the emerald leaves is the most brilliant crimson color I have ever seen. Perhaps it is because of its placement within so many other bright colors, or the way that the sun is causing its feathers to glow, but I can no longer remember if cardinals are as red as this. The Scarlet Tanager sings on and on as a soft breeze gently brushes his scarlet feathers, and for this moment he is indeed the most red creature in the whole world.

May 2, 2013. Tivoli Bays, Annandale on Hudson, New York

Great-crested Flycatcher
"Myiarchus crinitus"

"This is one of my favorite birds!" Susan exclaims excitedly as she shows me a photograph of a crested bird that is perched at the very top of a tree. "What kind of bird is this?" I ask her. "It's a Breep-o!" She responds. While this may not have been the most scientific name she could have given, it was definitely the most accurate. When she points out a Breep-o, or Great-crested Flycatcher (as the field guides like to call it) it is indeed croaking one thing and one thing only. "Breep! Breep! Breep!"

May 8, 2013. Tivoli Bays, Annandale on Hudson, New York

Red-eyed Vireo
"Vireo olivaceus"

"These birds actually drive me crazy after awhile. Once they get here, they stay all summer, and they never shut up." Susan explains her feelings on Red-eyed Vireos to me as we begin our walk around Thompson Pond. To me, a bird that sings constantly sounds great; if I can hear it, I can track it down and see it. It does not take us long before Susan hears one. "Here I am, in the tree, at the top, vi-re-o." She was right; its three-note chorus really does loop over and over. Which is fortunate, because the newly grown leaves are making it impossible to find the source of the singing. My neck is craned back so far it starts to grow sore. This bird really must be "at the top." But patience pays off, and finally, I notice a leaf that is a creamy color amongst the green. When I look at it through binoculars, I can see a white throat, gray cap, and, to my delight, the red rings in its eyes that give the chatterbox its name.

May 10, 2013. Thompson Pond, Pine Plains, New York

~You Came Back to Me~
May 6, 2013

So many things had to go just right for me to wind up at the train tracks at exactly 6:58 AM. I had to use up my extra snooze. I had to spend time debating whether or not to wear my jacket. I had to decide to wade through the water across the causeway even though it would freeze my feet, and I had to opt to walk out onto the tracks even though time constraints were telling me to turn back. But when I walk past the 97 sign at 6:58, the stars align. I can hear a birdsong that I cannot identify in the slightest. When I pursue the source, I find a small, grayish bird with a yellow wash hopping about in a brush. I stay far back at first, so as not to frighten it. I cannot figure out what it is, so I creep forward slowly, ever so slowly. The bird does not seem to mind. To my surprise, it bounces completely out of the brush and into full view, still singing merrily. Wheech, whee-choo! Over and over. I cannot believe how close I am getting. I still cannot figure out what the bird is...I am looking at it through the screen on my camera, and the strangest feeling comes over me, and at that same moment I can see the white eyes in its face. I feel the ground swaying beneath me, and my hands start to shake, because this is too impossible to be real. It is the White-eyed Vireo, it has come back; it is just the two of us on the train tracks as it sings. I have found it again, against all odds. My little bird has come back, and I am the one to find it. And now it is here, singing for me, and I am standing barely ten feet from it just watching and watching and knowing what it is and knowing that it has come back. In that small space between us there may have been no divide at all, and I cannot shake the feeling that surely, the White-eyed Vireo has come back for me. So many things had to go just right, and they did. And here we are.

Science says that animals cannot love in the same way that people can. Thus, to love anything other than a human is destined to result in love unrequited. But this White-eyed Vireo makes me wonder otherwise. There is a warmth spreading through my cold toes, my shaking hands, and my pounding heart; I cannot stop smiling, and I am all but skipping as I make my way back down the train tracks towards the swamp. I am so happy, deliriously happy; my little bird came back for me. Whether or not the vireo loves me does not matter. So much time and distance passed between us since I locked eyes with it for a split second on that murky day last July. This little bird came all the way back to where we first met and waited on the train tracks to sing just for me. It came back to me. And whatever this feeling is, I want to hold onto into it for as long as I possibly can.

~Stirrings~
May 8, 2013

GREAT CRESTED FLYCATCHER
"MYIARCHUS CRINITUS"

As we walk through the swamp under the gaze of the rising sun, the whichita! call of the common yellowthroat is everywhere, competing with the trilling of redstarts and the cawing of blackbirds. There are herons, sparrows, wrens, turtles, ducks, all swirling about in a milieu of mud and marsh and reeds. The swamp is alive, it is pulsing with energy and growth, and every footstep into the mire is a step that pulls me deeper into the swell. When we reach the tracks, the breep! of a Great-crested Flycatcher and the singing of a dazzlingly orange Baltimore Oriole steals the ear and eye. My White-eyed Vireo flies to greet us on tracks that go on forever.

If I could decide, right here and now, to step onto the worn wooden train tracks and just follow them into the endless horizon, would I? I can only imagine the wonders that lie beyond my world. What creatures are there to meet? What places are there to journey through? My existence is shaped by a gravel-covered road, a muddy causeway, bays filled with reeds and turtles and skies filled with soaring raptors.

It is a small world when compared to the one that I could find if I were to walk along the train tracks beyond the boundaries of Bardland...

BALTIMORE ORIOLE
"ICTERUS GALBULA"

COMMON YELLOWTHR
"GEOTHLYPIS TRICHA

Virginia Rail
" *Rallus limicola*"

Susan and I look for a Sora rail every time we go birding. We play its maniacal, whinnying laugh through her iPod and hold our breaths as we wait for a response. We never get one. But today, a few moments after we have given up waiting and have resumed our walk, we both freeze at the cackle that rises from the reeds below us. It is not a Sora. "Virginia Rail!" Susan is so excited that she is already bounding off the trail and down the hill. Elated, I follow her. This is one of those times when I feel so grateful that my middle-aged professor is so cool and more than willing to go crawling around a marsh looking for a bird. We pick our way as close to the water as we can. The cackling is growing louder. The bird must be here; it must be so close! I grab Susan's shoulder and point ecstatically at the shape emerging between the cracks in the reeds. It is dark gray, but led by a long pointy bill that is incredibly orange. As the reeds thin, I can make out the body of the chickenlike bird. I can barely breathe as it walks towards us. Its feet are hilarious, enormous clawed appendages that are bigger than its head. When Susan tells this story, she tells people that the Virginia Rail "practically walked into Christina's lap." And the wonderful thing is that this is not really an exaggeration.

May 10, 2013. Thompson Pond, Pine Plains, New York

Chestnut-sided Warbler
"*Dendroica pensylvanica*"

By May, the warblers have arrived in full force. The trees are crawling with color and exploding with song. I see a Chestnut-sided Warbler at Buttercup, all bright and twittery, but in the euphoric rush of seeing sixty-eight different birds in one morning, it becomes compressed into a single experience. Next spring, my ornithology class goes on a field trip to Bashakill State Park. It is warbler overload all over again, and my classmates and I can barely figure out where to point our binoculars first. Even the professor, Bruce, is giddy with excitement. There is something wonderfully warm about being with a group of people all genuinely interested in looking at birds. To me, it is even more amazing that they are all my friends. I love birding with these people; I love watching the expressions of wonder on the faces of the newly converted as they see a Black-throated Green or Blackburnian Warbler for the first time. Around mid-morning, we take a break to just sit (we woke up to catch the bus that left at 4:30 AM) I am filled with a sense of companionship as we revel in the birds we have already seen and talk excitedly about those we hope to find later. We watch in awe as a pair of Ospreys harasses a Bald Eagle out of their territory. Eastern Kingbirds, Red-winged Blackbirds and Yellow Warblers entertain us with their songs and antics. But for me, the bird that will forever bring me back to this day is the Chestnut-sided Warbler.

May 17, 2013. Buttercup Audubon Preserve, Stamfordville, New York

Yellow-billed Cuckoo
"Coccyzus americanus"

I heard a Yellow-billed Cuckoo long before I saw one. On the last birdwalk of my junior year at Bard, Susan, our friend Renee, and I head down Cruger Island Road altogether for the last time. Halfway down the road, Susan hears something unfamiliar. I hear it too. It is an echo-like, hollow popping sound uttered at a steady pace. Susan and I crash off the road into the woods to pursue it. It is maddeningly close, yet no matter how hard we bushwhack we cannot seem to find it. Renee hangs back, laughing at us. Out of desperation, Susan and I tape the bird. Its call is so distinct that we reason we can at least try to figure it out from the recording. A few days later, Susan emails me with a guess: a cuckoo. I listen to the sound of the Yellow-billed Cuckoo, and the calls match exactly.

For some reason or another, the morning I finally get to lay eyes on a Yellow-billed Cuckoo is the morning I go birding with Susan down Cruger Island Road for the last time as a Bard student. We know that road so well, she and I. I love it. And today, as always, it shows that it loves me back: it gives me a gift to commemorate this significant event in my life. Two birds that I swear could be flying bananas shoot over the clearing at the metal gate and land in a tree right above us. One immediately flies off, but the other sits and poses. I can see the black stripes on its long, thin tail and the crook in its yellow beak. It is such a funny-looking bird that I laugh aloud with both glee and gratitude.

May 21, 2013. Tivoli Bays, Annandale on Hudson, New York

Snowy Owl
"Bubo scandiacus"

Like a spirit refusing to split from its corporeal form, the ghostly owl clings to the skeletal branches of a treetop. It is an older male, made evident by the lack of barring across its soft chest. The biting wind that cuts across my face barely seems to phase the Snowy Owl. His feathers are blown askew; his perch careens back and forth. Yet the owl is king of his wintery kingdom. This is his world, his snowy, gray world. He is totally in control as his feet move from branch to branch. It is almost as though he is walking across the treetops. I look into his golden eyes- the only color within the snowstorm- and I feel my heart freeze as the white wraith finally leaves the skeleton and glides into the nothingness of the snowy dunes. He is out there somewhere, above the ocean capped with white plumes and the beaches that stream with snow and sand as they are pummeled with wind. It is though the Snowy Owl has melded into the elements themselves.

January 18, 2014. Salisbury Beach, Cape Ann, Massachusetts

"who cooks for you?"

64

~The Land Where only the Stars Light the Way~
February 3, 2013

It can be scary, to voluntarily venture into the night, to be swallowed by the dark. It is a surrendering of the senses as sight becomes useless and hearing questionable. It is hard to know what each sound is as every little cracking twig and rustling branch looms in the quiet.

It takes bravery to surrender oneself to the darkness. Or does it? Because really, what is there in the darkness to be afraid of? Monsters? Murderers? Not likely. Owls? Not a problem. There is nothing of be afraid of except the unknown, whatever it is that I create it to be. But in reality, the Tivoli Bays in darkness are anything but the unknown. I know this place. What lives in it, what the space looks like, feels like. I know that there is nothing here to be afraid of. So, I am not afraid. Instead, I am at peace. Because in this darkness, once the darkness is no longer a foe but a friend, there is beauty that cannot be found anywhere else. It can only be seen here, where nothing can be seen and only the stars light the way. And beneath those stars are memories, adventures, a safe place. Here, the stars shine like nowhere else in the world.

RED-SHOULDERED HAWK
"BUTEO LINEATUS"

Least Bittern
"Ixobrychus exilis"

The Tivoli Bays are one of the best places in New York to find Least Bitterns. Be that as it may, the odds of finding the tiniest and most secretive of North America's herons are still outrageously low. Their golden chests and black caps help them blend in perfectly with the phragmites and cattails of the North Bay. Susan and I are birding along the Cruger Island Road causeway when a tiny bird suddenly explodes from out of nowhere and dives into a bush just past the muddy walkway. "Least Bittern!" Susan calls out ecstatically. I strain to find the bittern, but to no avail. It seems to have vanished. Panicked, I crawl up onto the roots of the phragmites to try to get as close as possible. I am starting to worry that the first Least Bittern that has ever made an appearance in my life will disappear before I can really see it...but then just the slightest of movement, and I can finally follow the twisting branch to the neck of the Least Bittern and then, finally, to its golden eye. It is absolutely incredible how completely it can blend in. For all I know, there could be dozens of Least Bitterns all around me and I would never know. Susan and I agree that we should try to sneak closer to get a better look. Susan moves forward, camera at the ready- and I gasp with horror and hysteria as she sinks up to her thighs! I can almost hear the Least Bittern laughing with me as she grabs onto my arms and I somehow, miraculously, manage to help pull her out.

May 12, 2014. Tivoli Bays, Annandale on Hudson, New York

Cliff Swallow
"Petrochelidon pyrrhonota"

Cliff Swallows are my nemesis bird. I chased them down every single weekend last July, with no success. Weekend One: I head down to the railroad bridges that run parallel to the Hudson. Tree Swallows, Rough-winged Swallows, and Barn Swallows abound- but not a single Cliff Swallow. There is, however, a Department of Environmental Conservation officer waiting for me at the end of the Cruger Island Causeway which, unfortunately, has a big sign on it that says "No trespassing between January 1 and September 30." So attempt number one lands me with a trespass warning. Weekend Two: The least disastrous weekend. I bus-hop my way from Bard College to Beacon to look for Cliff Swallows on the docks of the waterfront. I must look at hundreds of swallows, including ones that have the characteristic white headlamp of the Cliff Swallow- but to no avail. Weekend Three: I bike twenty-seven miles East to Millbrook following the coordinates provided by my rival on eBird. It is ridiculously hot, and I spend over an hour and a half walking up and down the road only to realize that the coordinates are to a private residence. After weekend one, I am not about to put a single foot on private property. So I slowly pedal home, exhausted and overheated. Weekend Four: Little Brown Bats invade the house I am watching, and my mother insists I get rabies shots. So I bike for two hours to Norrie Point with a very painful rear end, only to find that the reported Cliff Swallows are actually Barn Swallows. So a year later, I am still looking for Cliff Swallows. I receive an eBird report telling me that there are a group of nests right under the eaves of the CVS in Saugerties. It seems almost too easy. Surely something will go wrong. But when my mom, Allison and I drive up to the CVS, sure enough, there are swallows flitting about everywhere. I feel a sense of pure triumph as a small blue head with a white headlamp pokes out of the muddy nest.

July 6, 2014. CVS, Saugerties, New York

Yellow Warbler
" Dendroica petechia"

The Yellow Warbler is the bird that changed everything. I was sitting in a canoe, trying desperately to spot what my professor, who was in the bow, insisted was a fist-sized bird called a yellow warbler which right in front of us. To my despair, it remained invisible- until I returned to shore, raced to the library, and opened a field guide. As my eyes finally fell upon the soft, golden feathers that cover the bird from head to tail, glowing amid the summer-green foliage in the background of the photograph, I could feel my chest contract with the awareness that I had stumbled upon something extraordinary.

Countless adventures after the day my world changed forever, I again sit in a boat trying to lay eyes upon birds. My professor and I shout the names of the creatures that flit, swim, and soar across the water ahead of us: "Tree swallow! Cormorant! Bald eagle!" The water is dizzyingly brilliant around us as the light dances atop the water like millions of diamonds. It is absolutely beautiful. Suddenly, I hear a birdsong that I will remember for as long as I live. "*Sweet sweet I'm so sweet!*" I turn my gaze to see a flash of gold not caused by the day's end. A smile stretches across my face as my eyes fall upon the small yellow orb hopping to and fro among the branches of a tree on the nearby shore. To me, its soft, gleaming feathers and its melodious voice are even more beautiful than the water's glittering surface. "Yellow warbler!" I call out.

April 20, 2013. Vassar College Farm, Poughkeepsie, New York

"All things are in motion, all is in process, nothing abides, nothing will ever change in this eternal moment. I'll be back before I'm fairly out of sight. Time to go."

— Edward Abbey,
    "Desert Solitaire"

www.ingramcontent.com/pod-product-compliance
Lightning Source LLC
Chambersburg PA
CBHW051025180526
45172CB00002B/467

9 781365 250347